EZStep Math presents:
Multiplication, the EZStep Way, for grades 3-7!

Hi, and welcome to the EZMath method of learning multiplication! Using prompts and step-by-step instruction, you will attain the knowledge and confidence to solve difficult multiplication problems with ease!

My philosophy of teaching math is to give students clear, step-by-step instruction, followed by adequate practice towards increased fluency!

Rules:

In this book, you will encounter many blank lines or spaces, on which you will write a single number, or digit. It's important to remember that you can only write one digit per blank. For example, you may write 5 or 2, like this, but you can't put more than one digit per blank place (75 is a no-no!).

The first problems we will learn will be one digit times two digits, without regrouping. Don't worry if you don't know what all of these words mean, as we will get back to them as we go along.

Numbers multiplied by each other are called multiplicands, and the answer is called the product. In this next problem, 3x32, 3 and 32 are the multiplicands.

Step #1:

32
x3
—

The 3 is the one digit number, and the 32 is the two digit number. Start in the ones column and multiply 3x2.

Does it equal 6?
Yes, of course it does! So, write a 6 in the blank space.

Step #2:

32
x3
_6

Now that you finished the one's place, you can multiply the ten's, or 3x3 (even though it's actually 3x30, because the upper three is in the ten's place). And the answer is......(drum roll).......

Quite right, 9!

Step #3:

32
x3
96

There you go!
3x2=6
3x30=90
So, 3x32=96

Does this make sense to you? How about we try another problem! Don't forget, only one digit per blank place!

Step #1:

```
  43
  x2
  ___
   _
```

2x3=

Step #2:

```
  43
  x2
  ___
   _6
```

2x4=
(really 2x40=)

Step #3:

```
  43
  x2
  ___
  86
```

2x43=86

4

Now, you get to fly solo, and try a few problems on your own! We'll go step-by-step.

Step #1:

24
x2
―

Step #2:

24
x2
_8

Step #3:

24
x2
48

Don't stop now, dudes and dudettes!

13	32	31	11
x3	x 2	x3	x4
39	_4		

23	44	23	34
x2	x2	x3	x2
_6			6_

14	18	22	33
x2	x1	x4	x3
_8			

Solutions for page # 6 :

```
  13        32        31        11
  x3        x2        x3        x4
  39        64        93        44



  23        44        23        34
  x2        x2        x3        x2
  46        88        69        68



  14        18        22        33
  x2        x1        x4        x3
  28        18        88        99
```

Are you ready for 1 digit times 3 digits? I think you are, so let's give it a shot! Just remember, we've added the hundreds place for one of the multiplicands.

So, it's 2x4, then 2x20, and finally 2x300, although we don't write more than one digit per blank place!

Step 1:   324
             x2
            ___
              _

Step 2:   324
             x2
            ___
             _8

Step 3:   324
             x2
            ___
            _48

Step 4:   324
             x2
            ___
            648

Give it a go!

```
 324        412        231
  x2         x2         x3
```

```
 411        214        322
  x2         x2         x3
```

```
 244        132        402
  x1         x3         x2
```

And solutions for page #9:

```
 324        412        231
  x2         x2         x3
 648        824        693


 411        214        322
  x2         x2         x3
 822        428        966


 244        132        402
  x1         x3         x2
 244        396        804
```

And more…….

```
 214        312        221
  x2         x3         x3


 414        628        232
  x2         x1         x2


 935        223        304
  x1         x3         x2
```

And more solutions:

```
 214        312        221
  x2         x3         x3
 428        936        663


 414        628        232
  x2         x1         x2
 828        628        464


 935        223        304
  x1         x3         x2
 935        669        608
```

Next thing: two digits times two digits! Yeah, let's get to it!

```
  T O     O=ones place
 xT O     T=tens place
```

How about some real, live numbers?

Step #1:

```
  24
 x12
  ―
```

2x4=?

Step #2:

```
  24
 x12
   8
```

8, good!

2x2=

Step #3:

```
  24
 x12
  48
```

4, right! We know it's really 40, as in 48.

Step #4:

```
   24
  x12
   48
    0
```

Uh oh, here comes trouble! Why is that **0** down there in the one's place? And what am I possibly supposed to write on that blank place? *Oh, help!*

Well, we finished multiplying **2x24=48**, right? Right!

Now we need to multiply **1x24**, although it's really **10x24**, because the **1** is in the **10's** place. So, we write the **0** there to hold the place. The second line is reserved for the **10's** place, and *always* needs to start on the right side with a **0**.

Step #5:

```
   24
  x12
   48
   40
```

1x4=4 (really 10x4=40). Next blank place is 1x2= (really 10x20=).

Step #6:

```
    24
   x12
    48
  +240
```

2x24=48
10x24=240

Now it's time to add the lines together!

14

Step #7:

```
   24
  x12
   48
 +240
  288
```

𝔓hew, that wasn't so bad, was it? And how about this weird font? I feel like I'm in the middle of a 𝔇racula book!

Hey, how do you like all of that empty space down there? Do you think you're done with the book? Sorry, but nooooooooooooooooooooooo! But you can draw something here, if you'd like!

EZStep right up (See what I did there? Get it? EZStep right up! Ha! Ho! Ha!) for a few practice problems!

Step #1:

  34
x2**2**

   —

2x4=?

Step #2:

34
x2**2**
 _8

2x3=?

Step #3:

 34
x**22**
 68

  —

Hold the place with what number?

Step #4:

34
x**22**
 68
 _0

2x4= (really 20x4)

16

Step #5:

```
  34
 x22
  68
 _80
```

2x3= (really 20x30)

Step #6:

```
   34
  x22
   68
 +680
```

𓆭𓆭 𓇗𓆸𓌞𓆰𓇥𓁝

Translation: Add them up!

Step #7:

```
   34
  x22
   68
 +680
  748
```

𓂀𓆭𓇥𓌞𓌞𓁝𓁝   (Translation: Yipee!!)

17

Okay, time to fly solo! We're going to wean you, little by little, from the tips, hints, pointers, and clues!

```
    43              13              22
   x21             x31             x41
   ___             ___             ___

  +   0           +   0           +   0
   ___             ___             ___
```

```
    12              21              31
   x23             x13             x22
   ___             ___             ___

  +   0           + ___           +   0
   ___             ___             ___
```

```
    14              23              33
   x12             x31             x21
   ___             ___             ___

  + ___           +   0           + ___
   ___             ___             ___
```

Solutions for Page #18:

```
   43          13          22
  x21         x31         x41
   43          13          22
 +860        +390        +880
  903         403         902


   32          21          31
  x23         x13         x22
   96          63          62
 +640        +210        +620
  736         273         682


   14          23          33
  x12         x31         x21
   28          23          33
 +140        +690        +660
  168         713         693
```

And more!

```
   15            33            42
  x11           x23           x12

  +  0          +  0          +  0
```

```
   32            43            23
  x22           x22           x13
```

```
   45            32            23
  x11           x31           x23
```

And solutions for page #20:

```
    15         33         42
   x11        x23        x12
    15         99         84
  +150       +660       +420
   165        759        504


    32         43         23
   x22        x22        x13
    64         86         69
  +640       +860       +230
   704        946        299


    45         32         23
   x11        x31        x23
    45         32         69
  +450       +960       +460
   495        992        529
```

Wow, you're progressing like a champ (not a chimp)! If you can do one digit times three digits, and two digits times two digits, without "regrouping", I think you're ready to take the next big step!

What happens when you multiply two numbers and the answer is more than one digit, and it doesn't fit in one blank place? Hmmmm, whatever can we do?

Let's take a look at the next problem and make some sense!

Step #1:

```
  15
  x5
```

You very well may know what the answer is already, but do you know how to solve it? The funny thing about math is that *knowing the answer isn't enough*! You need to know the process of how to solve it, so then you can solve any problem, not just the ones you already know or can do in your head easily.

Step #2:

15
x5
―

5x5=25, but we can't fit two digits in the one blank for each place value. The blank is in the one's column, so we write the 5, for five ones, on the blank.

Step #3:

―
15
x5
5

What might go on the blank above the 1 in the ten's column?

Step #4:

2
15
x5
5

25 is two tens and 5 ones, right? Right? I can't hear you! Shout it, "Right!"

Step #5:

```
  2
 15
x 5
 75
```

Huh? 7? Good gosh, where did that come from?

Let's see......5x5=25

Carry (regroup) the 2, which really is 20, and then add it in after figuring 5x1=5

Don't forget the 2!

5x1=5, then 5+2=7 (of course, it's actually 5x10=50, then add the 20 that you carried to make 70, because those numbers are in the ten's place!) A couple of more samples!

Sample 2.)

Step #1:

$$\overline{1}6$$
$$\underline{\times 3}$$

3x6=
___ goes in the one's column while ___ gets "carried" above in the ten's column.

Step #2:

1
16
$$\underline{\times 3}$$
$$\underline{8}$$

<u>8</u> goes in the one's column while <u>1</u> gets "carried" above in the ten's column.

Step #3:

1
16
$$\underline{\times 3}$$
48

Don't forget to add that regrouped 1 (really it's 10) after multiplying 3x1.

Sample 3.)

Step #1:

$\overline{2}3$
$\times 4$
―――

Step #2:

1
23
$\times 4$
―――
2

4x3=12
2 ones, and "carry" the 1 ten!
Have you heard this somewhere before? Is my voice getting old?

Step #3:

1
23
$\times 4$
―――
92

Don't forget to add that regrouped 1 (really it's 10) after multiplying 4x2.

Practice:

14 x4	16 x4	12 x6
18 x4	12 x5	12 x8
13 x6	15 x6	16 x5
25 x3	19 x4	15 x3
14 x7	19 x4	45 x2

Solutions:

```
  1          1          1
 14         16         12
 x4         x4         x6
 ──         ──         ──
 56         64         72


  3          1          1
 18         12         12
 x4         x5         x8
 ──         ──         ──
 72         60         96


  1          1          1
 13         15         16
 x6         x6         x5
 ──         ──         ──
 78         90         80


  1          1          1
 25         19         15
 x3         x4         x3
 ──         ──         ──
 75         76         45


  1          1          1
 14         19         45
 x7         x4         x2
 ──         ──         ──
 98         76         90
```

Practice:

26	13	18
x3	x7	x5

21	11	36
x4	x6	x2

23	28	23
x4	x3	x3

35	29	49
x2	x3	x2

24	18	25
x3	x4	x4

Solutions:

```
  1              1              1
 26             13             18
 x3             x7             x5
 ──             ──             ──
 78             91             90


                                1
 21             11             36
 x4             x6             x2
 ──             ──             ──
 84             66             72


  1              2
 23             28             23
 x4             x3             x3
 ──             ──             ──
 92             84             69


  1              2              1
 35             29             49
 x2             x3             x2
 ──             ──             ──
 70             87             98


  1              3              2
 24             18             25
 x3             x4             x4
 ──             ──             ───
 72             72            100
```

How about some hundreds? Btw, the blank space tells you where you need to write the regrouped tens or hundreds number.

```
  1
 ___              ___              ___
 145              372              349
x  2            x  3             x  2
____            ____             ____


 ___              ___              ___
 193              263              182
x  3            x  3             x  4
____            ____             ____


 ___              ___              ___
 227              492              326
x  4            x  2             x  4
____            ____             ____


___ (2 places)    ___              ___ (2)
 125              250              965
x  8            x  4             x  2
____            ____             ____
```

Solutions, coming up!

```
  1           2           1
145         272         349
x 2         x 3         x 2
───         ───         ───
290         816         698
```

```
  2           1           3
193         263         182
x 3         x 3         x 4
───         ───         ───
579         789         728
```

```
  2           1           2
227         492         226
x 4         x 2         x 4
───         ───         ───
908         984         904
```

```
 24  (2 places)   2          11
125              250        965
x 8              x 4        x2
────             ────       ────
1000             1000       1930
```

It's not 11! It's a **1** *each* in the ten's and hundred's places, meaning add 1 ten and 1 hundred in their columns!

And more practice:

```
  365        284        988
x   2      x   3      x   1
-----      -----      -----


  117        180        231
x   5      x   3      x   4
-----      -----      -----


  182        349        109
x   4      x   2      x   4
-----      -----      -----


  275        429        694
x   5      x   6      x   3
-----      -----      -----
```

And more solutions:

```
   1            21
 365           284            988
 x 2           x 3            x 1
 ---           ---            ---
 730           852            988


   3             2              1
 117           180            231
 x 5           x 3            x 4
 ---           ---            ---
 585           540            924


   3             1              3
 182           349            109
 x 4           x 2            x 4
 ---           ---            ---
 728           698            436


  32            15             21
 175           429            694
 x 5           x 6            x 3
 ---           ---            ---
 875          2574           2082
```

Our next problems are two digits times two digits with regrouping! In these problems you will be combining two multiplication skills you have learned so far, two digit multiplying and regrouping.

Step #1:

Hmmm.....*ya gotta* write the problem!

```
  34
x 25
```

Step #2:

```
  2
  34
x 25
───
   0
```

5x4=20
Write the 0 and "carry" the 2 to the tens place.

Step #3:

```
  2
  34
x 25
───
 170
```

5x3=15 (really 5x30=150)
15+2=17 (really 150+20=170)

Step #4:

```
   2
  34
 x25
 170
+  0
```

Hold the place with a 0 on the second line of the solution.

Step #5:

```
   2
  34
 x25
 170
+ 80
```

2x4=8 (really 20x4=80)
The carried 2 is from earlier, so we can ignore it.

Step #6:

```
   2
  34
 x25
 170
+680
```

2x3=6 (really 20x30=600)

Step #7:

```
   2
  34
 x25
 ───
   1
 170
+680
────
 850
```

Add 170+680, and don't forget to carry the 1 for the new group of 100 that you made in the hundreds column.

The answer is 850.

(Translation: What did you say? You would like some practice problems? Was that it? Your wish is my command! Let's go EZ step-by-step!!!)

Practice:

Step #1:

```
    4 6
x   2 5
―――――
```

5x6=

Step #2:

```
    ―
    4 6
x   2 5
―――――
    _
```

5x6=30
"Carry" the 3 (groups of ten)

Step #3:

```
  3
    4 6
x   2 5
―――――
  _ _ 0
```

5x4= (really 5x40, of course, even though you might be tired of hearing it!)

20+3=23

Step #4:

```
    3
   46
 x 25
  230
   ___
```

Hold the place....what number do you write here?

Step #5:

```
    ‾
    3
   46
 x 25
  230
   _0
```

2x6=   (yes, blah, blah, blah, I know it's really 20x6=)

Carry the ? above the 3!

Step #6:

```
    1
    3
   46
 x 25
  230
 + 20
```

Yeah!

Now, what's 2x4? Brace yourself: **really 20x40!!!**

Step #7:

```
    1
    3
   46
 x 25
  230
 +920
```

Okay, finish it off!

Step #8:

```
    1
    3
   46
 x 25
  230
 +920
 1150
```

Nice going!
¡Bien hecho!
Tres bien!
Hen hao!
Sehr gut!

Now................................go solo!

```
    33              71              42
  x 25           x  16           x  27
  _____          _____          _____

  +   0          +   0           +   0
  _____          _____          _____
```

```
    62              72              53
  x 32           x  34           x  23
  _____          _____          _____

  +              +
  _____          _____
```

```
    74              39              68
  x 32           x  12           x  19
  _____          _____          _____

  +              +
  _____          _____
```

Solutions:

```
    1
   33              71              1
 x 25            x 16             42
  165            426            x 27
+660            +710             294
  825           1136            +840
                                1134
```

```
   62              72              53
 x 32            x 34            x 23
  124            288             159
+1860           +2160          +1060
 1984           2448            1219
```

```
    1
    1               1              7
   74              39             68
 x 32            x 12           x 19
  148             78             612
+2220           +390           +680
 2368            468           1292
```

More problems-careful, they now have a digit in the hundreds place to multiply.

```
      324
x      23
─────────
```

```
      372
x      24
─────────
```

That's right, more!

```
      413
x      42
---------
```

```
      263
x      51
---------
```

Scary!!!

```
      758
x      65
---------
```

(You might want to check them!!!)

```
      1                    1
     324                   2                        412
   x  23                 372                      x  42
    ────                x  24                      ────
     972                ────                        824
  +6480                 1488                    +16480
   ─────               +7440                     ──────
    7452                ────                      17304
                        8928
```

```
                               34
      31                       24
     263                      758
   x  51                    x  65
    ────                     ────
     263                     3790
  +13150                   +45480
   ─────                    ─────
   13413                    49270
```

45

Phew, you've come a long way, and now the final challenge!! We have only one more type of problem to learn in this program!
Da, da, da............................

```
  564
x225
-----
```

How do you think we tackle *this* bad boy? Give it a shot!

Keep at it!

Try, try, and try again! (don't cry!)

Don't give up!

No peeking at the solution!

Unless you really need to!

```
       1
       1
      32
     564
    x225
    2820
   11280
 +112800
  126900
```

Did you see that one coming? Start with two 0's (00) on the third answer line, to hold both the ones and tens places.

And practice, naturally!

```
    461          913
   x272         x451
```

```
  455           823
x 226         x 512
```

And, out of the kindness of my heart, solutions!

```
        1
        4                           1
        1                           1
      461                         913
     x272                        x451
      922                         913
    32270                       45650
   +92200                     +365200
   125392                      411763
```

```
       11
       11
       33                          11
      455                         823
     x226                        x512
     2730                        1646
     9100                        8230
   +91000                     +411500
   102830                      421376
```

You're Done!! Congrats!! Take A Bow!!

www.ingramcontent.com/pod-product-compliance
Lightning Source LLC
Chambersburg PA
CBHW062233220526
45471CB00009B/3458